PROCEEDINGS

OF THE

WASHINGTON ACADEMY OF SCIENCES

VOL. III, PP. 487–506. NOVEMBER 7, 1901.

PAPERS FROM THE HOPKINS STANFORD GALAPAGOS EXPEDITION, 1898–1899.

IV.

ENTOMOLOGICAL RESULTS (4):

ORTHOPTERA.

[TEXT FIGURES 35–44.]

BY JEROME MCNEILL.

CONTENTS:

INTRODUCTION.

IN his paper on the Orthoptera of the Galapagos Islands[1] Scudder enumerates eight expeditions which have made collections there. These are Darwin in the *Beagle*, 1835; the Swedish frigate *Eugenie*, 1852; Louis Agassiz, in the *Hassler*, 1872; Commander Cookson, in the *Petrel*, 1875; Wolf, State Geolo-

[1] Bull. Mus. Com. Zool., Vol. xxv, No. 1, pp. 1–24, Pl. I–III, Cambridge, Sept., 1893.

gist of Ecuador, 1875; the *Albatross*, 1888; Alex. Agassiz, in the *Albatross*, 1891; and G. Baur for Clark University, 1891.

Scudder's paper is based on the specimens obtained by all the expeditions except the *Beagle*, *Eugenie* and *Petrel*, and he had seen specimens of all the species enumerated by him, except one, which is of uncertain occurrence. Butler referred two larval mantids to the genus *Vates* and Scudder concludes that it was impossible to have referred the specimens he knew to that genus.

Twenty species of Orthoptera are given in this paper, these species representing all the families except Phasmidæ. All the Blattids, however, five in number, are cosmopolitan species. Of the remaining *fifteen* two are of such doubtful occurrence that they may be excluded, leaving but *thirteen* indigenous species. Moreover, since the forficulid, *Anisolabis annulipes* Lucas, must be regarded as an introduced species, this number must be still further reduced to *twelve*. These twelve are included in eleven genera, representing four families. The Blattidæ and Forficulidæ are represented only by cosmopolitan species, and Phasmidæ is not known. Excluding the few doubtful cases, species are referred to thirty-six localities, or excluding the cosmopolitan forms to twenty-eight.

The above outline of our previous knowledge of Galapagos Orthoptera has been given in order to clearly show the additions made by the Hopkins Stanford Expedition.

Messrs. Snodgrass and Heller spent about six months in the Islands, usually staying only a few days at each one. Albemarle was however twice visited and several weeks spent in its exploration. No special effort was made to collect Orthoptera as the purpose of the expedition was to make as complete collections as possible of the entire fauna and flora of the islands. The insect collecting was done chiefly by Mr. Snodgrass.

The following table exhibits the distribution of the Orthoptera.

It will be seen from this that Orthoptera were obtained from all islands previously visited except two, Jarvis and Gardner, and from four new localities, Narboro, Culpepper, Abingdon and Bindloe. Omitting the cosmopolitan Blattids, which he made no effort to collect, Mr. Snodgrass obtained all the species hitherto known, except two, *Closteridea bauri* Scudd. and

DISTRIBUTION OF ORTHOPTERA IN THE GALAPAGOS ISLANDS.

NOTE. New species in black face type. Localities previously reported are shown by an *, not previously reported, by a †.

Species.	Tower.	Albemarle.	James.	Jarvis.	Duncan.	Indefatigable.	Barrington.	Chatham.	Hood.	Gardner.	Charles.	Narboro.	Culpepper.	Abingdon.	Bindleo.
1. Anissolabis annulipes	...	†	...	...	...	...	...	*†	...	...	...	...	...	...	...
2. **Temnopteryx snodgrassii**	...	†	...	...	...	...	...	...	...	...	...	...	...	...	...
3. Periplaneta americana	...	...	...	...	...	...	...	*	...	...	*	...	...	...	...
4. Periplaneta australasiæ	...	...	...	...	...	...	...	†	...	...	*	...	...	...	...
5. Nauphæta cinerea	...	...	...	...	...	...	...	*	...	...	...	...	...	...	...
6. Nauphæta circumvagans	...	...	...	...	...	...	...	...	...	...	*	...	...	...	...
7. Leucophæa surinamensis	...	...	...	...	...	...	...	*†	...	...	*	...	...	...	...
8. Galapagia solitaria	...	*†	...	...	...	†	...	...	†	...	...	...	...	...	...
9. Closteridea bauri	...	...	...	...	...	...	...	*	...	...	...	...	...	...	...
10. Sphingonotus fusco-irroratus	...	†	?	...	...	†	†	†*	†	*	*†	...	...	...	...
11. Schistocerca melanocera	...	*†	*†	*	*†	*†	*†	*†	...	...	†*	†	...	†	†
12. Schistocerca literosa	†*	...	...	...	...	...	...	*†	†*	...	*	...	...	...	...
13. Halmenus robustus	...	...	*	...	...	†*	...	...	...	...	...	...	...	...	...
14. **Halmenus cuspidatus**	...	†	...	...	...	...	...	...	...	...	...	...	...	...	...
15. **Halmenus choristopterus**	...	...	...	...	...	...	...	...	...	...	†	...	...	...	...
16. Desmopleura concinna	...	...	*	...	...	...	...	...	...	...	...	...	...	...	...
17. Anaulocomera darwinii	...	...	†	...	...	†*	...	*	...	...	...	...	...	...	...
18. **Liparoscelis cooksoni**	...	*†	†	...	...	†*	...	...	†	...	*†	...	...	...	...
19. **Liparoscelis palludicola**	...	†	...	...	...	...	...	...	...	...	...	...	...	...	...
20. Liparoscelis pallidus	...	...	...	...	...	...	†	...	...	...	...	...	...	...	...
21. Conocephalus insulanus	...	...	...	...	...	...	...	*†	...	...	...	...	...	...	...
22. **Xiphidium exitiosum**	...	...	†	...	...	†	...	...	...	...	...	...	...	...	...
23. Gryllus galapageius	†	*†	...	...	...	...	...	*?	†	...	?	†	†	†	...
24. **Nemobius speculi**	...	†	...	...	...	...	...	...	...	...	...	...	...	...	...
25. Cycloptilum erraticum	...	†	...	...	...	...	...	...	†	...	*	...	...	†	...
26. **Cycloptilum lepismoide**	...	†	...	...	...	...	...	...	...	...	...	†	...	...	...

Desmopleura concinna Scudd. In addition to this, he obtained eight new species or two-thirds as many (indigenous species) as had been previously known, besides several subspecies and races not included in the table. These species fall in six genera, three of which had not been previously reported from the Islands. Of these genera one, *Temnopteryx*, belongs to the Blattidæ, a family not hitherto represented in the indigenous fauna. These specimens were obtained in fifty-four localities or excluding the cosmopolitan species in fifty. Thirty-two of these localities, or excluding the cosmopolitan species thirty, are new.

Fifteen of the Galapagos Islands are now known to possess Orthoptera. This number does not include small islets (like

South Seymour) lying so near larger ones that they may be regarded as forming an integral part of them. The number of known species is twenty-six. This number includes six species which are almost certainly cosmopolitan, but excludes the doubtful *Vates* and *Gryllus* given in Scudder's List.

The Hopkins Stanford Expedition, as is obvious, has greatly enlarged our knowledge of the orthopterous fauna of these Islands. But it has done more. By collecting a large series of individuals of the several species it has shown how potent a factor isolation is in the production of species. This example is especially instructive because the differentiation cannot well be explained by natural selection, or by the effect of environment. All the indigenous species are apparently new. Of these 20 indigenous species twelve are incapable of flight. Of the remaining eight, one, *G. solitaria*, has a wingless female; another, *Gryllus galapageius*, is wingless twice as often as winged. Thus nearly seventy per cent. of the native orthoptera are wingless.

Of the native species nine, including the cockroach, mantis, crickets, and the *Liparoscelis* were in all probability brought to the islands on timber or in partially decayed logs, the Mantid, at least, in the egg state. Though nothing is known of *Halmenus*, yet judging from some of its allies, this species may have been introduced in the egg state, the eggs being deposited in partly decayed or at least dead timber. The *Schistocera* and the *Sphingonotus* were almost certainly brought there by their own wings, and it seems probable that the long-winged *Desmopleura* was introduced in the same way. As to the four remaining species there is greater uncertainty. Both *Anaulocomera* and *Conocephalus* have efficient wings but neither are great flyers and it seems very unlikely that they should have been able to keep on the wing long enough to reach the islands. Nor can we readily suppose that they could have drifted in tree tops, since their wings would have tempted them to fly, probably thus ruining their only chance for safety. All things considered, the most reasonable hypothesis is that they were introduced in the egg state, the eggs being attached to twigs or leaves.

To account for the presence of the last two species presents the greatest difficulty of all. Nothing is known, I believe of the habits of *Closteridæ*. Only one species of the *Truxalinæ*, *Chlöealtis conspersa* Harr., so far as I know, has been observed to lay its eggs in wood, and it is possible that either *Closteridea* has, or its ancestors had, the same habit. All the *Xiphidia*, I believe, live on the ground and never on trees, so that its occurrence here is altogether unlikely and I can form no plausible theory of the manner of its introduction. The problem is not simplified by the fact that it is found on two islands, James and Indefatigable.

Albemarle is represented by the greatest number of species (twelve), and this is quite to be expected, since it is the largest of the islands and its greatest length lies across the path of the currents and trade winds. It is on the leeward side of the group. Indefatigable is next in size and, with Chatham, it has the next largest number (seven) of native species. Chatham is fourth or fifth in size and is, I believe, the only one of the islands now inhabited. It is the easternmost of the group. Charles and James have each six representative species. James is probably fourth in size and has no more than its proportionate number of species. Charles is smaller and its relatively larger number of species is probably due to the fact that it has been inhabited and presumably more frequently visited than most of the others. It is difficult to understand why Hood, which is rather smaller than Charles, and has been much less frequently visited, should have the same number of species. The remaining islands, with one exception, are small and three of them—Abingdon, Bindloe and Tower—lie far from the others, so that the small number of species credited to each is natural. The exception referred to is, however, curious.

Narboro ranks second or third in size. It has but three species credited to it and all these are found on Albemarle. This condition of things may be partly explained by the fact that the island is completely cut off from the prevailing currents by Albemarle, so that it is apparently impossible that drift wood should ever reach it from the main land.

All the specimens of orthoptera collected by Mr. Snodgrass

now form a part of the entomological collection of Stanford University, and the types of the new species here described are consequently to be found there.

FORFICULIDÆ.

ANISOLABIS ANNULIPES Lucas.

Forficasila annulipes LUCAS, Am. Soc. Ent. Fr. Bull., 1847, LXXXIV.
Anisolabis annulipes BOR., Biol. Centr. Am. Orth., p. 5.
Anisolabis bormansi SCUDD., Bull. Mus. Comp. Zool., Vol. XXV, p. 5.
Anisolabis maritima? BRUN. (*nec* Bor.), Proc. U. S. Nat. Mus., Vol. XII, p. 192.

Careful examination of numerous specimens has convinced me that Scudder's species is identical with a Pacific coast species, *A. annulipes*. The joints of the antennæ vary in what appear to be full-grown specimens (those 4 or 5 mm. in length have from six to eight joints) from thirteen to nineteen. In many cases, no doubt, the terminal joints are lost. The position of the ring of luteous is quite uncertain. There may be one, two or three joints beyond it and it may include a single joint or as many as three, or it may be absent entirely. A variable number from one to four or more of the basal joints are also luteous. Scudder's description was based upon a single female, consequently his description of the forceps would apply to that sex only. In the male these appendages are very unsymmetrical, the left one being decidedly but regularly curved throughout its length, while the right is bent, rather than curved, not far from the middle, and sometimes the apical half is nearly at right angles to the basal. The femora are only very obscurely banded with fuscous; indeed, in many cases this character is indistinguishable. The specimens from Albemarle are much larger than the others, the largest measuring 20 mm. in length. The same species was found at Clipperton Island, perhaps a thousand miles northwest of the Galapagos group. Snodgrass collected specimens from Albemarle, Chatham and Clipperton. Since all these islands are inhabited, and as this is a cosmopolitan species found in both the new and old world, and as no specimens have been found on any of the uninhabited islands, there can be little doubt that it is an introduced species.

Clipperton, 22 specimens varying from 4 mm. to full grown; Chatham, 4 specimens; Albemarle, 13 specimens, 5 males.

BLATTIDÆ.

Subfamily BLATTINÆ.

TEMNOPTERYX SNODGRASSII sp. nov.

Description.—Dark chestnut, somewhat lighter on the pronotum and tegmina. Labrum, coxæ and femora testaceous; tibiæ and feet chestnut. Pronotum parabolic, the posterior edge barely arcuate, horny, polished and remotely and shallowly punctate, with the anterior and lateral margins a little lighter than the disk. Tegmina short, not quite covering the first segment of the abdomen, horny, with the veins exceedingly obscure and the anal sulcus entirely obsolete; sutural and costal margins straight and parallel; posterior margin straight and slightly oblique, with the exterior angle moderately rounded. Anterior femora with three spines near the middle of the anterior lower carina, followed by a series of minute spinules on the apical half, no spines on the posterior lower carina; terminal spines $\frac{2}{2}$. Middle and posterior femora with three spines on the anterior lower carina and four on the posterior lower carina; terminal spines $\frac{2}{1}$. Supra-anal plate triangular and slightly keeled (male), or transverse and deeply incised (female). Cerci twice as long as the supra-anal, plate in the median line, somewhat depressed, dark chestnut; stylets of the male very small.

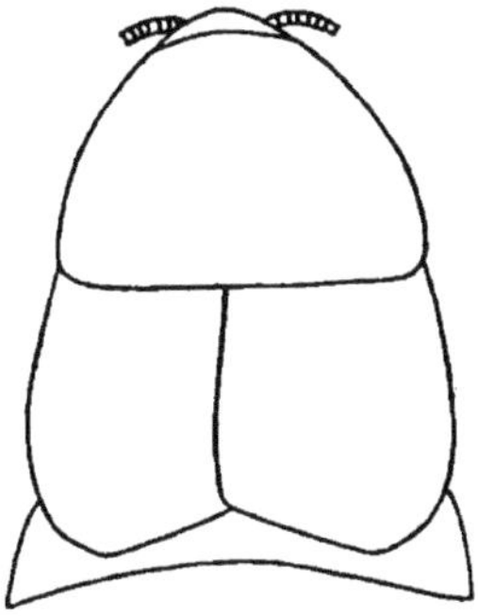

FIG. 35. *Temnopteryx snodgrassii*, female, head, pronotum and anterior part of abdomen, dorsal view. (×8.)

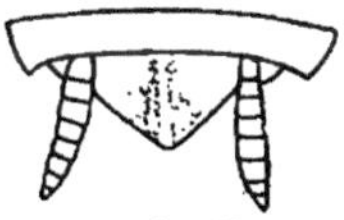

FIG. 36. *Temnopteryx snodgrassii*, female, dorsal view of end of the abdomen. (×8.)

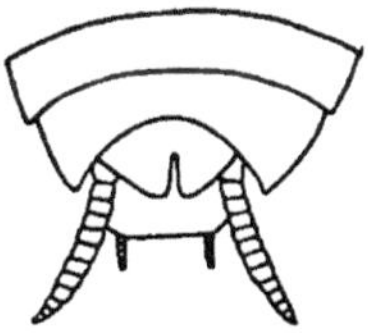

FIG. 37. *Temnopteryx snodgrassii*, male, dorsal view of end of abdomen. (×8.)

Measurements.—Length of body, ♀ 8 mm., ♂ 7 mm.; tegmina, ♀ 3, ♂ 3; pronotum, ♀ 3, ♂ 3.

This is the only blattid which is probably indigenous. All the other species in the collection belonging to this family and the three previously reported are well-known cosmopolitan forms.

This species is perhaps more closely related to *Temnopteryx sumichrasti* Lucas than to any other, but it is readily distinguished from it by the plain color and by the divided supra-anal plate of the male.

The specimens upon which the species is based were collected at an elevation of 2,000 feet, on the south side of Iguana Cove Mountain, Albemarle Island, on very wet ground covered with an abundant and luxurious vegetation of ferns, shrubs and vines. The specimens were found under leaves. As Mr. Snodgrass searched diligently for insects living under stones and leaves, it is highly probable that the species is restricted to this single locality. One of the specimens contains, partly extruded from its body, a species of *Gordias (?)*.

Albemarle, 1 male, 3 nymphs.

Subfamily PERIPLANETINÆ.

PERIPLANETA AUSTRALASIA Fabr.

Previously recorded by Bruner from Charles Island. No effort was made to collect blattids, as they were supposed to be cosmopolitan species entirely.

Chatham: 1 male.

Subfamily PANCHLORINÆ.

LEUCOPHÆA SURINAMENSIS Linn.

Two immature specimens from Chatham, both females.

Three previous expeditions to the Galapagos have reported this species from Charles and Chatham.

MANTIDÆ.

Subfamily MANTINÆ.

GALAPAGIA SOLITARIA Scudd.

The collection contains eleven specimens from three localities. Seven of them are from Albemarle, the island from which came the single male and female on which Scudder's description is based. These should, therefore, be typical specimens but they are all dull brown in color instead of clay-yellow, the females usually having the pronotum and especially the abdomen with three longitudinal fuscous stripes on an obscure testaceous ground. The males, however, and especially the single adult, are grayish-brown with the tegmina of the same color but with the cross-veins on either side of the main veins deep fuscous. The specimens from Indefatigable are so different in size and color that they deserve to rank as a distinct subspecies. The adults are very light testaceous with the cross-veins of the tegmina not at all infuscated. Both forms have a very marked depression on the inner side of the anterior femora for the reception of the tibial spur, and beyond this impression three large spines. The Albemarle form, which may be called *typicalis*, has distad to this impression four spines on the outer and eight on the inner side; the Indefatigable form, which may be called *major*, has five and ten spines respectively.

Length of body, *typicalis*, 40 mm., *major*, 41 mm.; tegmina, *typicalis*, 25 mm., *major*, 28 mm.; hind femora, *typicalis*, 12.5 mm., *major*, 18 mm.

Albemarle, 1 male, 2 females, 4 nymphs; Indefatigable (South Seymour Island), 2 males, 1 female; Hood, 1 nymph.

ACRIDIDÆ.

If we exclude the doubtful *Vates* and *Gryllus*, the only two indigenous species previously described from the Galapagos which are not found in the present collection belong to the family Acrididæ. These species are *Closteridea bauri* Scudd., a truxalid, and *Desmopleura concinna* Scudd., an acridid. The four other previously known species were collected in large numbers and Mr. Snodgrass has prepared a paper dealing with their variations and distribution.

Subfamily ŒDIPODINÆ.

SPHINGONOTUS FUSCO-IRRORATUS Stål.

Numerous specimens from six islands, viz.: Charles, 11 males, 28 females; Albemarle, 4 males, 9 females; Barrington, 1 male, 8 females; Hood, 3 males, 7 females; Chatham, 4 males, 14 females; Indefatigable, 13 males, 31 females.

Subfamily ACRIDINÆ.

SCHISTOCERCA MELANOCERA Stål.

Represented by numerous specimens which, as Mr. Snodgrass has shown, are separated into distinct races which generally but not always have their range limited by the geographical division of an island.

Specimens were collected on Charles, Albemarle, Barrington, Indefatigable (South Seymour), James, Narboro, Duncan, Abingdon, and Bindloe.

SCHISTOCERCA LITEROSA Walk.

Taken on Chatham, Tower, and Hood.

HALMENUS ROBUSTUS Scudd.

In addition to this species heretofore recognized in the specimens from Indefatigable, there are, on James, two other quite distinct forms which I am inclined to consider of specific value.

Indefatigable, 2 males, 6 females.

HALMENUS CUSPIDATUS Snodgrass (MS.).

This form is represented by so many specimens that there seems little doubt that the species is a good one.

Albemarle, 2 males, 5 females.

HALMENUS CHORISTOPTERUS Snodgrass (MS.).

This form is unfortunately represented in the collection by only a single specimen.

Charles, 1 female.

LOCUSTIDÆ.

Subfamily PHANEROPTERINÆ.

ANAULOCOMERA DARWINII Scudd.

The collection contains a single pair which agree well with the published description, except that the anterior and middle femora have no trace of spines on the lower surface.

James, 1 female. Indefatigable, 1 male.

LIPAROSCELIS Stål.

Nesæcia SCUDD., Bull. Mus. Comp. Zool., Vol. XXV, 1893, No. 1, p. 20.

Thirty-seven specimens of this genus are in the collection, taken from six of the islands, which are, however, very unequally represented. Twenty-six specimens are credited to Albemarle while Indefatigable has but a single larva, and Barrington and James a single adult each. So great is the difference among these specimens that it is necessary to recognize three species and one subspecies, or probably four species, and each of the islands represented possesses at least one distinct variety. All of these forms are, without doubt, more closely related to each other than to other species of the genus, and all are easily separated from their congeners. In 1895, Brunner, in his Monographie der Pseudophylliden, assigned four species to this genus, two being from Mexico and one from Colombia. *Liparoscelis cooksoni* is at once distinguished from two of these species by its smooth face. It is probably much more closely related to the remaining species, *L. nigrispinis* Stål, but is readily distinguished by the fewer spines on the under side of the fore and middle femora and by the acute unemarginate subgenital plate and straight ovipositor of the female.

Albemarle, 26; Indefatigable, 1 larva; Barrington, 1; James, 1.

LIPAROSCELIS COOKSONI Butler.

Agræcia cooksoni BUTL., Proc. Zool. Soc. Lond., 1877, pp. 87–88.
Bucrates? cocoannus? BRUN. (*nec* Boliv.), Proc. U. S. Nat. Mus., Vol. XII, p. 192.
Nesæcia cooksoni SCUDD., Bull. Mus. Comp. Zool., Vol. XXV, 1893, No. 1, p. 20.
Liparoscelis cooksoni BRUNN., Monag. Pseudophyl., p. 176, 1895.

Butler's type was from Charles, so there is no doubt as to the forms to which his name applies. True, he had some immature specimens from Albemarle, and some or all of these may have belonged to another species. Brunner's specimen was from Charles and its identification may be considered certain. As Scudder's specimens (except a single immature one from Indefatigable) came from Albemarle, they probably belong here, though they are somewhat larger than my specimens from the same locality, and the ovipositor is more than proportionally longer; howsever, this may be due to the fact that the method of measuring this organ used by Scudder and myself are different. He measures a straight line from the tip to the junction of the upperand lower laminæ where they leave the body; in other words he measured the chord and I the arc. If the *curvature* was measured, and especially, in this case, if the keel of the lower lamina was measured, the length here given would be increased two or three millimeters.

FIG. 38. *Liparoscelis cooksoni* Butler, female, ovipositor and end of abdomen.

Brunner gives Scudder's measurements and they are exactly the same, except that he adds the length of the four femora, so that his specimens were the same as Scudder's. His description does not fit the Albemarle specimens in all respects, but on the whole it suits this species better than any other.

To this species are referred the specimens from Charles Island, a part of those from Albemarle, and those from James and Hood. The specimens from Hood, however, form a distinct subspecies and, as before stated, each of the islands probably has its distinct race which cannot be properly characterized here for want of material. The Charles race, however, is decidedly larger and more robust than the others; the tegmina overlap plainly in the females; the spines of the under side of the femora are 2 or 3, 2, 4, and the spinules on the outer carina of the upper side of the hind tibiæ are 2 or 3. The Albemarle specimens are decidedly smaller; the tegmina of the females scarcely more than touch at the base on the basal half; the femoral spines are 3 or 2, 2 or 3, 4 or 5; the hind tibiæ spinules (outer carinæ upper side) 4 or 5 or none beyond the middle. The James form is represented by a male only, and it must be compared with the male of the Albemarle form as there is no male from Charles. Compared with this form, the male has the spinules on the outer carinæ of the upper side of the hind tibiæ uncertain in number and so exceedingly minute as to be scarcely discoverable. The musical organ occupies about one-third instead of one-half the length of the tegmina, and the stylets of the subgenital plate much exceed (by their whole length and a part of the plate equal to half their length) the cerci. These forms resemble each other in having the pronotum strongly tuberculate, the tegmina of the female decidedly shorter and little if any longer than the pronotum, the spines of the hind femora being generally four, the spinules of the outer carinæ of the hind tibiæ never more than five and confined to the distal half, and the ovipositor comparatively short and wide and quite strongly curved.

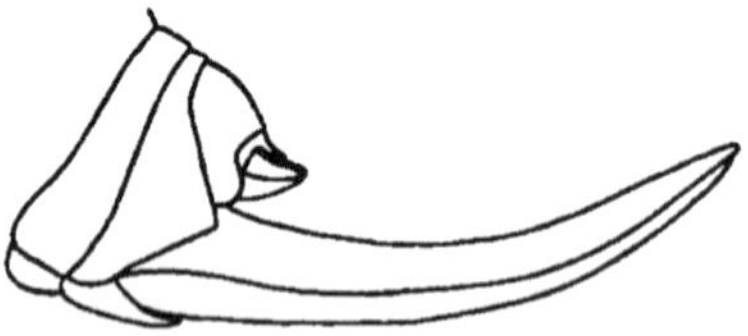

FIG. 39. *Liparoscelis cooksoni ensifer*, female, ovipositor. (×3.)

The remaining form is of subspecific value and may be known as *Liparoscelis cooksoni ensifer* (Fig. 39) from its comparatively longand gracefully curved ovipositor. It differs from *L. cooksoni* Butler in the following particulars: It is relatively more slender and more compressed; disk of the pronotum only obscurely tuberculate and moderately rugose; tegmina of the female at least as long as those of the male, decidedly longer than the pronotum; femoral spines, 3, 3, 5; spines of the outer carinæ of the upper side of the hind tibiæ eight, occupying the distal three-fourths; cerci of the male short, exceeded by the subgenital plate (exclusive of the stylets) by at least their own. length; ovipositor of the female long, decidedly curved and slender.

Hood, 1 male, 2 females, May 13 and 14.

Measurements are: Length of body, ♂ 34 mm., ♀ 33; pronotum, ♂ 6.5, ♀ 7; tegmina, ♂ 9, ♀ 9; hind femora, ♂ 17, ♀ 18; length of ovipositor, ♀ 14.2; width of ovipositor, 2.

Charles, 2 females, 6 nymphs, May 9.

Measurements of female: Length of body, 37.5 mm.; pronotum, 7 to 8; tegmina, 7; hind femora, 18.5 to 19; ovipositor, 11.5 to 12.

Albemarle, 1 male, 2 females, June 9 and 12.

Measurements.—Length of body, ♂ 27 to 28 mm. ♀ 25 to 28; pronotum, ♂ 6.2 ♀ 6.5; tegmina, ♂ 8 to 8.7, ♀ 5.5; hind femora, ♂ 13, ♀ 14; length of ovipositor, ♀ 10 to 10.5; width of ovipositor, 2.5.

James, 1 male, 1 larva.

Measurements of male: Length of body, 28 mm.; pronotum, 6; tegmina, 8; hind femora, 14.5.

Indefatigable, one larva.

LIPAROSCELIS PALUDICOLA sp. nov.

Description.—Color nearly uniform chestnut brown except the sides and top of the head, which are grayish and the face an hour-glass-shape and the disk of the pronotum which are light clay color. The

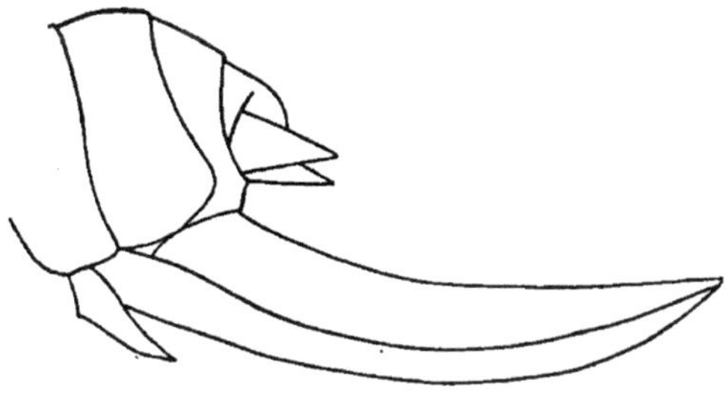

FIG. 40. *Liparoscelis paludicola*, female, ovipositor. (×3.)

pronotum is obscurely tuberculate on the disk, with a distinct median sulcus which is colored like the body but fades out on the front edge of the prozone and on the posterior part of the metazone. The tegmina of the male are but slightly longer than those of the female; in both they are much longer than the pronotum, and oblong in shape, with the apex rounded. The femoral spines are usually 3–2–4; the spines on the outer carinæ, upper side of the tibiæ about three confined to the distal half. The cerci are not exceeded by the stylets of the subgenital plate. The ovipositor is very stout and moderately curved.

This species is closely related to *L. cooksoni* but easily distinguished by its peculiar coloration as well as by its size and several structural characters. According to Mr. Snodgrass, it is confined to mangrove swamps, where it is found under the bark of *Avicennia*, a tree which always occurs with the mangrove. Specimens were numerous at Tagus Cove and a single one was found in a mangrove swamp at Elizabeth Bay. The species is sluggish in its habits, and its song was not heard.

Measurements.—Length of body, ♂ 40 mm., ♀ 42; pronotum, ♂ 6, ♀ 6; tegmina, ♂ 10–12, ♀ 10; posterior femora, ♂ 14, ♀ 19; length of ovipositor, 16; width of ovipositor, 3.

Albemarle, 12 males, 6 females.

LIPAROSCELIS PALLIDUS sp. nov.

A single specimen, rescued from a mocking-bird, is so different from all the others that, although it is without a head or forelegs, or middle legs, I feel obliged to make it the type of a new species. The following is a description:

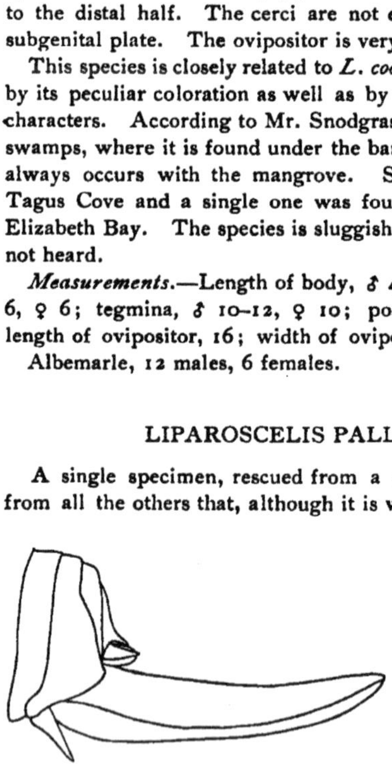

FIG. 41. *Liparoscelis pallidus*, female, ovipositor. (×3.)

Description.—Yellowish-white, body and limbs mottled with purple, becoming black on the pronotum; ovipositor whitish on the basal, black on the distal half. Pronotum evenly and rather strongly tuberculate, the two transverse sulci distinct, the median sulcus apparent only at the principal transverse sulcus; shoulders somewhat distinct on the metazone, obsolete on the prozone; posterior margin straight; lateral lobes with the posterior margin very slightly sinuate and distinctly less oblique than the anterior margin. Tegmina ovate, subacuminate, touching only at the base, with the median and radial veins distinct and the dorsal angularly separated from the lateral field. Hind femora armed below on the distal half with seven spines, the first two minute; hind tibiæ armed above with three spinules on the outer and nine on the inner carinæ. Ovipositor only slightly curved on the upper, more decidedly on the lower margin; subgenital plate triangular acute.

Measurements.—Length of body (estimated), 37 mm.; pronotum, 6.5; tegmina, 10; hind femora, 17.5; length of ovipositor, 14; width of ovipositor, 2.5.

Barrington, 1 female.

Subfamily CONOCEPHALINÆ.

CONOCEPHALUS INSULANUS Scudd.

It is rather remarkable that this species is known only from Chatham. Mr. Snodgrass reports that it was common in a wet pasture about 1,000 feet above sea level. It was not found elsewhere. Three of the males and one female nymph are brown. In this variety the tegmina are maculate pretty evenly with obscure fuscous. Besides the fuscous stripe which borders the stridulating organ in both varieties, the brown specimens have the upper half of the lateral lobes of the pronotum infuscated, and this color is continued as a stripe on the tegmina bordering the radial vein. The lower half of the lateral lobes is also infuscated more or less, and this region is separated from the upper part by a moderately distinct stripe of the lighter ground color.

Chatham, 5 males, 3 females, 4 nymphs.

XIPHIDIUM EXITIOSUM sp. nov.

Description.—Color, green with the usual dorsal stripe on the head and pronotum. Pronotum widens posteriorly and continues to the hind margin; it is darkest medianly and on the edges. The vertex is only slightly pinched at the eyes and is scarcely wider than the basal joint of the antennæ. The posterior border of the lateral lobes of the pronotum is scarcely perceptibly sinuate. The posterior femora are nearly as long as the body, unarmed beneath. The ovipositor is at least three-fourths the length of the hind femora and is very slightly curved.

This species is closely related to *X. nemorale* Scudd., from which it differs strikingly in color. It is also readily distinguishable by the

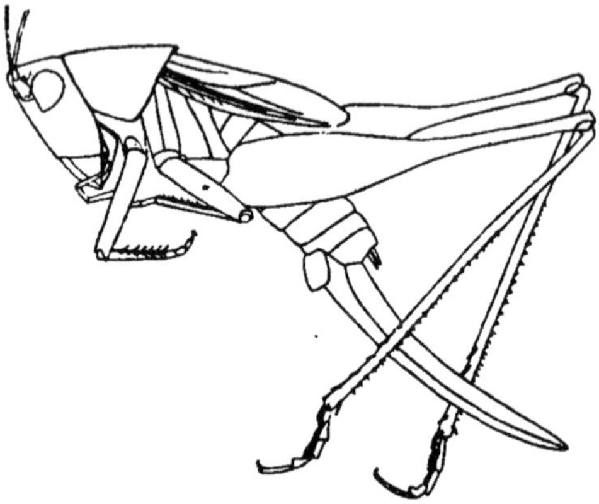

FIG. 42. *Xiphidium exitiosum*, female, lateral view. (×3.)

length of the tegmina which are much shorter in the male and much longer in the female, and by the longer and straighter ovipositor.

Measurements.—Length of body, ♂ 15 mm., ♀ 17; tegmina, ♂ 6, ♀ 8; hind femora, ♂ 12, ♀ 16; length of ovipositor, 12.

Indefatigable, 1 female, 2 nymphs; James, 2 males, 2 females, 1 nymph.

GRILLIDÆ.

Subfamily GRILLINÆ.

GRYLLUS GALAPAGEIUS Scudd.

This species is represented in the collection by a considerable series taken in seven of the islands. The variation in size is considerable as indicated by the measurements, and the color variation (especially among the immature) is quite marked. Less than one-third of the adults (three males and three females) are winged. The obscure, minute, reddish spots are always present in the females, but are generally entirely wanting in the male. On the other hand, while the females show considerable uniformity in the color of the tegmina, these are entirely black in males from Albemarle and Narboro, and wholly testaceous in a single male from Culpepper. The variation in color among the immature specimens is much more marked, but I believe they all belong to the same species. Five specimens from Culpepper and one from Abingdon are testaceous mottled with fuscous. Two specimens from Tower and one from Hood have the general body color, which is black, considerably varied with testaceous on the pronotum and abdomen. Eight other specimens from Hood and two from Albemarle are entirely black, except for the hind femora which are always either plainly or obscurely transversely barred with fuscous on a testaceous ground, and sometimes, also, a pair of minute reddish spots on the disk of the pronotum and very obscure testaceous or reddish spots on the abdomen and pronotum. Scudder was unable to distinguish the tympanum on the inner side of the anterior tibiæ. It is present in all these specimens though minute.

The examination of 42 specimens of Grylli of all ages from seven of the islands compels the conclusion that the second species of this genus, which Scudder thought to be indicated by three immature specimens from Charles, probably does not exist; at the same time the absence of all specimens of this genus from Charles leaves the matter in doubt.

Measurements.—Length of body, ♂ 14 to 23 mm., ♀ 18 to 22; tegmina, ♂ 7.5 to 14, ♀ 10 to 14.5; hind femora, ♂ 10 to 14, ♀ 14; ovipositor, 18 to 19.

Of its song, Mr. Snodgrass says it "consists of a continuous chirping sound differing from the intermittent sound made by the small tree crickets (*Cycloptilum*) found in swamps."

Albemarle, 3 males, 2 nymphs; Hood, 3 males, 8 females, 9 nymphs; Chatham, 2 males, 2 females, 1 nymph; Tower, 2 nymphs; Abingdon, 1 nymph; Narboro, 1 male; Culpepper, 1 male, 6 nymphs; uncertain, 1 female.

NEMOBIUS SPECULI sp. nov.

Description.—Dark brown more or less varied with testaceous. Sometimes the lighter color predominates but usually the darker, except in the legs which are always obscure testaceous banded more or less regularly with brown. The face, occiput and disk of the pronotum are obscure testaceous maculate with brown; on the posterior part of the disk the markings are minute round spots; the lateral lobes of the pronotum are brown with about seven round light spots, five above and two below; the veins of the tegmina are for the most part testaceous. The frontal costa is rather thickly pilose with stiff black hairs. The antennæ are about four times as long as the body. The pronotum is only slightly narrowed anteriorly, less than one and a half times as wide as long and usually sparsely pilose with long bristly hairs. Tegmina of the male about two-thirds as long as the abdomen, truncate or very slightly arcuate posteriorly with the external and internal posterior angles extended into rather distinct spines; speculum very large, including nearly all of the dorsal field, the apical field therefore very much reduced; wings apparently obsolete. The anterior and middle femora are strongly compressed, the latter decidedly arcuate. The posterior tibiæ are armed with three spines on both the inner and outer angles; these spines are situated on the apical half, the lowermost inner one being half as long as the tibiæ and about as long as the metatarsus; spurs of the posterior tibiæ, three, on both the inner and outer side, the upper inner one being about as long as the metatarsus, the latter a little less than twice as long as the two joints of the tarsus. Cerci about as long as the head and pronotum. Ovipositor of the female of uncertain length, not toothed at the tip.

FIG. 43. *Nemobius speculi*, male, dorsal view except hind legs. (×8.)

Measurements.—Length of body, 5 mm.; tegmina, 2.5; hind femora, 4.

The collection contains about twenty-three specimens of *Nemobius*, all collected at the same time in a single locality. Unfortunately only two of them are mature, and both are males, though a considerable number of both males and females are in next to the last moult. Since the tables for determining species in this large genus (at least forty-five species are described) are generally based on the females, and since in many cases the males are not known, I have not been able to satisfactorily determine the position of this species; but since all species indigenous to the Gala-

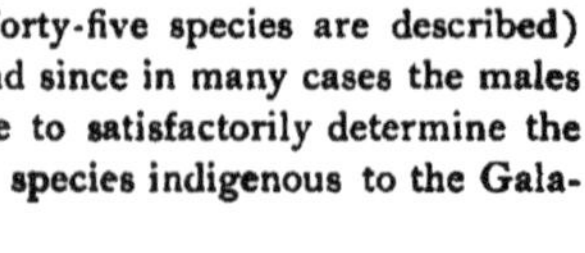

FIG. 44. *Nemobius speculi*, male, posterior tibia. (×8.)

pagos have proved to be new, the following description is given without much fear of its specific distinctness.

Albemarle, Tagus Cove, 2 males, 20 nymphs.

Subfamily MYRMECOPHILINÆ.

CYCLOPTILUM ERRATICUM Scudd.

This species has been known heretofore by a single specimen, a male from Charles Island. The present collection contains five adults and several nymphs. The female is so unlike the male in appearance as to make it difficult to recognize it as belonging to the same species. The best preserved specimens are dull cinnamon brown, especially anteriorly, becoming chestnut posteriorly, with the legs sometimes lighter. The surface of the body and legs is smooth or very sparsely pilose. The female is wingless with the pronotum subquadrate, very moderately widening posteriorly, and not covering the mesothorax. The ovipositor is as long as the hind femora. A few scales are found on the under side of the body and sometimes a single row bordering some of the abdominal segments. The antennæ are about twice as long as the whole body.

It occurs in mangrove swamps on the under side of leaves and under the bark of dead trees, and is reported to be numerous, but shy and difficult to catch. Mr. Snodgrass says: "Their singing can be heard the whole day in the swamp. It consists of an intermittent chirping sound, the intervals between the series of chirps being short, however, so that the whole has a continuous sound, but it is readily distinguished from the louder and continuous chirping of the large cricket living everywhere beneath the rocks." Two of the specimens from Albemarle, apparently young males, though the pronotum is very small, have large tegmina and wing pods.

Albemarle, 3 males, 1 female, 5 nymphs; Hood, 1 male.

CYCLOPTILUM LEPISMOIDE sp. nov.

This species is remarkable among Orthoptera for the possession of scales. Though this character is shared by other species of the genus *Cycloptilum*, it is much less evident. *C. lepismoide* differs from *C. erraticum* in its color, which is very dull brown, becoming blackish posteriorly. The legs are quite thickly pilose, and these, together with the under surfaces, are luteous or white varied with black. The white and black of the under surface and the black of the head is due to the presence of thickly imbricated scales. These scales are of various forms but possess the common characters of a prominent carina, which ends distally in a point, and a complicated sculpture. This sculpturing consists of thickened plates which are arranged in rows parallel, approximately, to the median carina, and also in rows forming a right angle with the first set. Under a moderate power of the microscope the appearance presented is that of concentric curved depressions running lengthwise of the scale, and less evident depressions curving radially from the midrib. The legs, and especially the posterior pair, are plainly shorter and more robust. The last abdominal segment of the male is shaped like the bow of a boat and from its apex extends a pair of minute forked appendages which are approximate nearly throughout their length. The anal cerci are quite as long as and distinctly stouter than in *C. erraticum* and they are quite thickly pilose with hairs, the longest of which are more than twice as long as the diameter of the cerci at the base. On account of these hairs the cerci appear much stouter.

Measurements.—Length of body, ♂ 6.5 mm., ♀ 9; pronotum ♂ 3.5, ♀ 2; hind femora, ♂ 3.5, ♀ 5.

Except for the omission of any mention of scales, Scudder's description of *Cycloptilum erraticum* would appear almost equally well to each of these species. It seems unlikely that he has overlooked so striking a character as this, however, and the smaller size of the only adult male of *C. lepismoide* has determined me to consider this one as the undescribed form.

Found on the leaves of bushes and possibly also on the leaves of the mangrove.

Albemarle, 1 male, 1 nymph; Narboro, 6 nymphs; Abingdon, 1 female.

Milton Keynes UK
Ingram Content Group UK Ltd.
UKHW021942101123
432363UK00004B/136

9 781020 562273